Teresa Lurdes Miguel Monjane

Changing methodologies for better maths learning

AF294837

Teresa Lurdes Miguel Monjane

Changing methodologies for better maths learning

ScienciaScripts

Imprint
Any brand names and product names mentioned in this book are subject to trademark, brand or patent protection and are trademarks or registered trademarks of their respective holders. The use of brand names, product names, common names, trade names, product descriptions etc. even without a particular marking in this work is in no way to be construed to mean that such names may be regarded as unrestricted in respect of trademark and brand protection legislation and could thus be used by anyone.

Cover image: www.ingimage.com

This book is a translation from the original published under ISBN 978-613-9-71848-1.

Publisher:
Sciencia Scripts
is a trademark of
Dodo Books Indian Ocean Ltd. and OmniScriptum S.R.L publishing group

120 High Road, East Finchley, London, N2 9ED, United Kingdom
Str. Armeneasca 28/1, office 1, Chisinau MD-2012, Republic of Moldova, Europe
Printed at: see last page
ISBN: 978-620-7-71592-3

Dedication

To the memory of my father Cuamba Gueliche Miguel, my mother Maria Cristina Alexandre Muteque, my siblings Filipe António Cuamba, Mário Jorge Cuamba and Catarina Amália Guilamba, all deceased, may their souls rest in peace.

To my husband Tomás Arone Mondlane, my daughters Albena Tomás Mondlane, Tânia de Lurdes Tomás Mondlane, Mutatela Cristina Tomás Mondlane, Jéssica Teresa Tomás Mondlane, Andreia Cecilia Tomás Mondlane and Tomásia de Tomás Arone Mondlane. And to my siblings, Júlio Cuamba, Maria João Miguel and Maria Leonor Guilamba.

Thank you

Behind the finished work, whatever it may be, you have to look at the people who made it possible, directly or indirectly, for it to be realised. "Saying thank you" is a mark of honesty that allows us to associate the institutions and, through them, the people who contributed, each in their own way, to the author's work.

This work owes a great deal to Professor Jo António Capece & Professor Vinício De Macedo Santos, for their rigorous guidance, careful and patient reading of the manuscripts and preliminary versions, and to the teachers, coordinators of the Maths subject and students from the General Education Schools.

This work would certainly not have seen the light of day without the support of my family. My thanks go first to God for giving me life, to my husband, my daughters for their moral support and particularly to my brother, Professor Dr Mário Jorge Cuamba, my mother Maria Cristina Alexandre Muteque and my sister Catarina Amália for their encouragement in my studies,

(Peace to their souls).

Summary

This article entitled: "Mathematical Modelling in the Constructivist Approach as an Alternative to Changing Teaching Practices in the Classroom in the 1st Cycle of General Secondary Education" aims to analyse the potential of the connection between Mathematical Modelling and constructivism and its contribution as a methodology in pedagogical practices in the classroom in the teaching-learning process in General Secondary Education (1st Cycle) at Matola Secondary School. This study came about as a result of the realisation that the pedagogical practices in place in the majority of schools do not respond to current social demands, as students are limited to receiving the knowledge transmitted by the teacher without criticising it, which compromises the quality of education and their skills. In view of this, the use of mathematical modelling combined with the constructivist approach contributes to an effective teaching practice, making students active, reflective, participative and capable of producing their own knowledge. This article was produced using the extensive works of various authors such as: Bassanezi (2002); Barbosa (2007); Imenes (1989), Lopes and Borba (1994), Gazetta (1989) among others who deal with Mathematics Education, mathematical modelling and constructivism. For its production, a qualitative and quantitative methodology of a descriptive nature was used. Therefore, taking the above assumptions as a basis, among the different strands of constructivism, the perspective that adopts mathematical modelling as an environment and as a strategy in mathematics classes, could provide a fertile field of possibilities aimed at training and teaching practices that invest in the student's creative and innovative capacity, thus contributing to the construction of diverse knowledge. We therefore see the possibility of an approach to teaching mathematics that combines the assumptions of mathematical modelling and Piagetian constructivism with great potential for innovation in teaching practices, given the presence of a common denominator between these two perspectives, since they make it possible to focus on problem situations from everyday life and the students' social context, bringing their cognitive experience and understanding into play for meaningful mathematical learning.

Keywords: Mathematical Modelling, Constructivism, Mathematical Teaching-Learning Process

Index

Introduction

Education is a tool that aims to form a human being capable of solving problems and transforming society through the knowledge acquired. Throughout her professional career, the author has found that traditionalist teaching based on the transmission and memorisation of knowledge does not seem sustainable, as it does not allow students to use their creativity to understand and solve didactic problems in the classroom, a fact that contributes to their low performance in the subject of Mathematics in particular, contrary to the plans, objectives and strategies outlined in the Government's Five-Year Programme (PQG, 20052009), in which the new political, economic and socio-cultural context at national and world level calls for a new vision of the ESG.

The problem that led to this report is that the process of teaching and learning mathematics in Mozambican schools is still characterised by the traditional, expository or deductive model, in which the teacher presents concepts, principles, deductions or statements from which conclusions or consequences are drawn, and which is imposing, characterised by education based on the imposition of knowledge and norms to be respected, contrary to the objectives set out in the National Education System (PCESG), where the curriculum places the student as an active subject, with learning geared towards the development of life skills. As a result, teaching continues to be the reproduction of pre-established behaviours in a system of reference that mutes the students, making them incapable of elaborating and expressing their critical thinking, thus limiting them to passively accepting the authority of the teachers. So to what extent can mathematical modelling and constructivism work together to contribute as an alternative to changing teaching practices in the classroom? And how?

Current trends in ESG in the southern region, on the African continent and around the world almost all point to the integral development of the student, equity and the improvement of teaching quality through a realistic, relevant and professionalising curriculum that can lead the student to autonomous learning. It is on the basis of these current trends that teaching practices in the teaching and learning process in the classroom must contribute to improving the quality of teaching, to solving today's challenges through a diversified and professionalising curriculum, promoting cognitive development in the student, values and life skills. Changing the way we teach and learn in the classroom is more important than any structural manipulation of the contents of the curriculum, which is unfortunately the reality in our schools. However, the proposal to combine mathematical modelling and constructivism, as a fundamental tool for innovations and conventional teaching methods in the classroom, can go some way to forming an active subject capable of constructing their own knowledge and able to develop skills and abilities in solving problems related to life. In fact, the successful implementation of the new competency-based education system (knowing, doing, being and being) depends fundamentally on the teacher (continuous training), the student (interest in learning), school managers, education agents (supervision) and everyone involved in the process.

In view of the above approaches, promoting maths learning remains a major challenge to be overcome in the classroom. Good quality teaching and learning, aimed at understanding mathematical ideas, requires a break with traditional practices based on memorising and repeating content. To this end, it is necessary to review teaching approaches and strategies and try to adopt differentiated methodologies that value problematisation, that are close to the student's interest and curiosity, that recognise their competence by encouraging them to mobilise their own knowledge, that have room for questions and answers, that consider error and difficulty as elements that are part of the teaching and learning process and are therefore not insurmountable barriers. Attention to these aspects can help improve the quality of students' maths learning.

However, discussing issues that are aimed at promoting changes in pedagogical practices in the classroom, given that school failure in maths is still high, means, in a way, reflecting on new teaching and learning methodologies.

However, the connection between mathematical modelling and constructivism as a conventional teaching method in the classroom, on the one hand, can help students to reflect on the realities of everyday life, i.e. extramathematics relating to mathematics and, on the other hand, as a fundamental instrument for innovation contributing, in a way, to the formation of an active subject capable of constructing their own knowledge, capable of developing skills and abilities in problem solving. This connection, as a methodological alternative for teaching mathematics in the ESG, is a challenging strategy, because it needs to break down the barriers of traditional teaching towards student-centred teaching, where the student has an active role in constructing their own knowledge and preparing proactive professionals in relation to autonomous and meaningful teaching and learning.

1. Historical aspects of the National Education System in Mozambique (SNE)

With regard to the historical aspects of Mozambique's National Education System, it can be said that it is the reflection or result of a process characterised by four major periods. Let's take a look:

The first period begins in the colonial era and lasts until 1975 (the year of Mozambique's national independence). This period was characterised by an education system restricted to a very narrow stratum, defined in cultural and racial terms. The system was conceived as an elitist education. The educational conceptions of this period continued to exert great influence on the central parameters of the education system in subsequent periods.

The second period actually began with national independence in 1975 and was characterised by a major effort to extend education to all Mozambican citizens. However, this process was interrupted by the war of destabilisation (the so-called 16 Years War), which shook the Mozambican state from 1976 onwards. As a result of this war, many teachers lost their lives, students were kidnapped and infrastructure was destroyed, especially in the 1980s.

The third period began with the General Peace Agreement and after the country's first multi-party general elections in 1994. This is considered to be a period of social stability and economic growth, also characterised by a return of investment in the education sector.

The fourth period began in 2000 and continues to the present day, with constant reformulations of the curriculum. In the government's programme from 2000 to 2004, education was integrated into "social development", which also included health, employment, culture, sport and the promotion of women's participation in economic and social life. With regard to the education sector, the government set itself the general objective, according to Castiano & Ngoenha (2013, p. 125). To widen the population's access to formal education, with a particular focus on basic education, without neglecting general secondary education, professional technical education, higher education, and other non-formal educational frameworks.

In fact, education is the driving force behind social development. The focus of this education was the formation of the "New Man". Basic secondary education is conceived as fundamental in preparing the man of tomorrow. In fact, this was Samora Machel's central thought after independence:

"to train the new man to respond to the demands of the new Mozambican society that was being built. Therefore, education is the main pillar aimed at forming citizens capable of transforming their personal and social lives."

We can therefore see that the transformations and innovations in the curriculum were aimed at improving the quality of education. In fact, the main challenge of the current curriculum is to make teaching more relevant and of higher quality. Bearing this principle in mind:

The aim is that, on completing basic education, the graduate will have acquired the knowledge, skills and values that will enable them to integrate effectively into their community and society in general. It is up to basic education to train students who are able to reflect and be creative, i.e. capable of questioning reality in order to intervene in it for their own benefit and that of their community (INDE/MINED, 2003: 21).

It can be seen in the above excerpt that all the efforts made were and still are aimed at forming a conscious, reflective, critical citizen capable of transforming and questioning personal and social reality, for the good of themselves and the community in which they find themselves. However, the results have not yet been achieved.

It has become clear that in order to make the objectives more achievable, it is essential that the education system, and in particular the teaching of maths, incorporate changes in training, teaching practices, approaches and methodologies in order to help students overcome their learning difficulties, improve their understanding of mathematical concepts and improve their performance.

Mathematical modelling is a methodological resource, although it is not the only possible approach, it represents an alternative methodology with the potential to promote significant changes in students' mathematical learning. Obviously, modelling can be an alternative to established school practices, as it has favoured students' learning and mastery of mathematical content, unlike traditional teaching - in which the mathematics teaching-learning process is flawed, compromising students' cognitive development.

According to BLUM et al (2007), interest in mathematical modelling in education has been growing both internationally and in Brazil.

There is ample evidence of the flexibility of this methodology in transforming teaching practices in the classroom and in the interaction of students' prior knowledge of social reality with academic mathematical content, as well as in everyday problem situations.

In Mozambique, as in other countries around the world, maths plays a fundamental and indispensable role in the integral formation of students. This idea is supported by a set of official curricular guidelines currently in force in the country, among which we highlight the Objectives of the Strategic Education Plan (PEE, 2006-2011):

- To contribute to improving the quality of teaching by providing students with relevant learning that is appropriate to the country's socio-economic context;
- Responding to the challenges of today through a diversified, flexible and professionalising curriculum;

e,

- Broadening the universe of choices, training young people both for further study and for the labour market and self-employment.

In the research carried out by the Ministry of Higher Education, Science and Technology (MESCT, 2004), one of the problems highlighted about ESG refers to *"a weak culture of science and technology in secondary schools"*. The findings of this research are summarised in some of the results mentioned here as follows:

- Highly academic, theoretical teaching without, however, giving priority to practical skills that can enable graduates to easily enter the labour market;
- Learning is fundamentally based on memorising procedures. Understanding concepts and developing skills through observation, visualisation, experimentation, deduction and generalisation do not seem to be part of teaching practice in the classroom.

Hence the need to think about some changes in the process of teaching and learning maths in general secondary education, since the reality of the national education system in Mozambique (SNE, law 6/92) has pointed out that the objectives of education are the formation of an active and participatory young person, endowed with skills, values, abilities and competences for work.Achieving the goals set for the learning of students in general secondary education (ESG), in the subject of maths, presupposes the combination of various factors, but the full commitment of the teacher and the active participation of the student are particularly indispensable in the teaching and learning process. There is therefore a complementary relationship between the pedagogical practices of teachers and the committed participation of students in the teaching-learning process in all subjects in general and in maths in particular. It is therefore necessary for the teacher to adopt a teaching method that stimulates the student's creativity and contributes to the development of the ability to work independently.

In the teaching and learning process, we believe that the constructivist perspectives of mathematical modelling are based on the student's creative capacity and on teaching practices that contribute to the construction of diverse mathematical knowledge. Meanwhile, through mathematical modelling, the teacher, in his or her classes, starts from the day-to-day knowledge that the students have in order to arrive at academic mathematics. This process facilitates the understanding of mathematical content, lessons are student-centred and, because they are student-centred, students are more active, participative and innovative.

This enables students to move from theory to practice and solve everyday problems.

However, it is on the basis of guidelines such as those mentioned above that maths teaching is seen as a means of enabling students to develop both logical thinking skills (reasoning) and to use mathematical knowledge in its practical dimension and to solve everyday problems.

The modern world points to the need to adapt maths to a new reality, based on problem-solving, exploring everyday situations, thus showing students the need and importance of learning maths, guaranteeing interdisciplinarity as a fundamental role in the student's integral formation and transversality as a way of discovering strategies for using maths to build specific concepts, attitudes, skills and abilities, with the aim of forming active, participative and visionary citizens.

For Mozambique, as for any other country with similar economic, social and cultural characteristics, it is an ongoing challenge to train qualified teachers capable of managing quality teaching so that students succeed in the teaching-learning process, particularly in the subject of maths.

In line with the highly theoretical (traditional) teaching that predominates in Mozambique, the main focus of which is often on fulfilling the vast programmes of the local curriculum, without giving space to the ultimate objectives, which are to open up horizons to create skills in the student, promoting their training and preparation to act critically and autonomously in the society in which they find themselves. In this context, the connection between mathematical modelling and constructivism is being proposed as an alternative for pedagogical practices in the classroom, with the aim of creating an enthusiasm and taste for mathematics in teaching and learning, and to stop thinking of it as a *"seven-headed beast", in other words,* conceiving of mathematics as an area of knowledge that is very difficult to understand, assimilate and master. That's why few people study maths. This is not true; maths is an area of knowledge that is accessible to everyone, just like any other scientific area.

Thus, taking the above into account, this research is based on the idea that the articulation between Mathematical Modelling and Constructivism can be an effective teaching methodology for the subject of Mathematics, since this perspective enables a closer approach to everyday problems, allowing the active participation of the student and promoting meaningful learning.

2. Challenges in mathematical learning

The great challenges of globalisation demand education that promotes quality. Mozambique's development depends on the existence of properly trained individuals, endowed with the competences set out in the new General Secondary Education programmes, which involve understanding a set of knowledge, skills, attitudes and values necessary for life, enabling students to face the reality of working in an increasingly modern and competitive economy. The realities mentioned above have spurred interest in applying a proposed methodology for teaching maths. In this sense, we can consider modelling in the constructivist approach as a methodological tool for this, according to Bassanezi,

"Mathematical modelling is the art of transforming problems from reality into mathematical problems and solving them by interpreting their solutions in the language of the real world" Bassanezi" (2002:16)

According to Bassanezi, Mathematical Modelling, as a possible alternative in teaching and learning, seeks, based on the problems that arise from a topic addressed in the classroom, a model that is suitable for acquiring solutions, based on an everyday situation and consistent with its application, Contrary to the traditional method in which the student is merely a receiver of a set of theories and formulas, without knowing where and how to apply them in their relationship with the real world, as well as admitting the hypothesis of the same situation occurring in the teacher himself, while in the constructivist approach the student develops activities that will allow him to build his own knowledge, because, according to Piaget ,

"Constructivism is a psychological theory of learning in which the pupil builds new knowledge on an existing cognitive structure in his mind, thus bringing about a qualitative cognitive change". Piaget (1969:14)

Therefore, from Piaget's perspective, the construction of new knowledge in the student is based on a set of processes, schematised within the nervous system, susceptible to continuous change, interacting with the environment, going through three phases: adaptation, assimilation and accommodation. According to Wadsworth (1996), these schemas consist of mental or cognitive structures through which the individual intellectually adapts to and organises the environment. It is from this perspective that this thesis is written, i.e. to problematise the issues, difficulties and impasses observed in the teaching of mathematics in the context of school education in Mozambique and to identify principles, guidelines and possibilities for approaches to mathematics in the classroom that will subsidise training processes and teaching practices with a view to improving the quality of teaching in this area. In order to carry out this research, and taking as a reference part of the extensive international bibliographic production available today in the field of Mathematics Education, we start from the assumption that by taking as a basis, among the different strands of constructivism, the perspective that adopts mathematical modelling as an environment and as a strategy in Mathematics classes, we have a fertile field of possibilities aimed at training and teaching practices that invest in the student's creative capacity and contribute to the construction of diverse knowledge.

The teaching-learning process is very complex, but in fact the fundamental problem is to respond to the challenges currently imposed in education and in the Curriculum Plan for General Secondary Education in Mozambique by promoting and developing the learning of mathematics, based on the use of mathematical modelling in a constructivist approach in order to increase the success rate and reduce the level of failures in mathematical learning, as well as creating student interest and participation in the classroom, contributing to the construction of their own knowledge in the learning of mathematics.

However, the traditional method is still predominant in the classroom, with the teacher at the centre of the lesson, explaining the content, drawing up exercises to fix it and finally assessing it, while the student remains a mere reproducer of the definitions and mechanically receives ready-made formulas, They don't have the ability to argue, let alone raise doubts. They lack the ability to reason, because they aren't led to reason based on everyday situations, nor are they led to draw their own conclusions by problematising the content they have learnt, thus not allowing for reflective learning without responding to the needs of today's society, contrary to the objectives defined in the PCESG. Its full realisation implies the combination of a variety of factors, including teacher and student motivation, the effectiveness of the curriculum, the teaching methods adopted, access to teaching materials, infrastructure and many others.

According to BASSANEZI (201:2008),

"Mathematical modelling, as a teaching-learning process in training or specialisation programmes, presupposes a course plan with well-defined objectives and guided by basic guidelines, such as: "

'A To enable teachers to change the concept of educational practice by freeing them from certain myths regarding the use of calculators, mathematical rigour, subject sequencing, assessment, etc...;

'A Develop motivations for innovative actions that awaken creativity;

'A Valuing mathematical knowledge in a global context and its power to act in specific situations;

'A Valuing available human resources, exploring and developing the talent of trainee-educators so that they feel able to contribute to the community in which they work;

'A Keep interdisciplinarity in mind, combining maths with other sciences so that it serves as a tool for understanding and possibly changing reality;

'A Interrelating experimental and theoretical factors, i.e. not losing sight of the very essence of the "mathematical attitude".

'A Taking into account the specific realities of each region and the interests of students, with the aim of increasing their motivation and effective participation in the wider community or environment of which they are part as citizens.

This doesn't mean adopting the popular thesis that "science in a developing country must be regional" - which would be a mistake since science either seeks universal explanations based on observable data, or it isn't science.

In this context, it can be seen that there is a need to invest in training and motivating the teacher as a mediator in order to tackle the new techniques and strategies used in the classroom as a central component for the success of the whole teaching-learning process. Modelling is currently a scientific method widely used in natural sciences such as physics, astrophysics, chemistry and biology, whose advances in terms of research can be seen in recent decades.

3. Curriculum Dimensions

3.1. Maths content" dimension
The CMESG must be structured according to content-based standards: (i) numbers and operations; (ii) algebra; (iii) geometry; (iv) systems of units of measurement; and (v) data analysis.

3.2. Alternative conceptions" dimension
Alternative conceptions are a set of knowledge (ideas and explanations) or experiences that diverge from scientific concepts, which learners bring into the classroom.

Generally, these conceptions are subscribed to by teachers and are resistant to extinction through traditional or conventional teaching strategies.

The teaching-learning process (TLP) techniques for remedying alternative conceptions are based on:

- -Cognitive conflict is the demonstration of dichrepancy between alternative conceptions and scientific conceptions.

- -Analogy is the creation of bridges between alternative conceptions and scientific conceptions.

- -Mathematical modelling is the manipulation of objects in the environment or dynamic models on the computer to solve problems.

3.3. Maths-Technology-Society (MTS) Dimension
The MTS perspective in the EAP has the role of preparing learners to be able to face the changing world.

For this reason, the CMESG must observe the approach: (i) based on the relevance of the topic; (ii) transdisciplinary; (iii) historical; (iv) philosophical; (v) sociological; and (vi) problematic.

In addition, the CMESG must observe the category of: (i) motivation through MTS content; (ii) casual introduction of MTS content; (iii) systematic introduction of MTS content; (iv) maths based on MTS content.

3.4. Problem-solving" dimension
The ESMC should be structured according to process-based standards in problem solving:

- -The improvement of knowledge and the stimulation of new learning;

- -The development of mathematical language, reasoning and critical thinking.

- -The use of mathematical models to communicate mathematical ideas and to recognise connections between ideas in contexts outside mathematics.

- -For this reason, solving maths problems should include small demonstrations and small investigations, as well as simple exercises and problems.

3.5. Practical work" dimension
- -S Practical maths work consists of manipulating objects in the environment, measuring

instruments or dynamic models using computers.

-*S* The purpose of practical maths work is to introduce new ideas.

3.6. Professional standards

-The valorisation of socio-cultural and ethno-mathematical aspects in the approach to mathematics.

-Alternative conceptions as stimuli for the construction of knowledge.

Problem-solving and practical work to develop skills, abilities and creativity.

4. On Education and Maths Education

At this point, we will try to develop a conceptualisation of education in the *broad* sense of the term and mathematical education in its narrow sense. Education is one of the problems that man has faced and continues to face ever since he existed.

For education is conceived as an extremely important tool for the intellectual and moral formation and integration of man into society.

Therefore, it is education that provides us with the types of values that man must cultivate, learn, assimilate and, above all, live within society. That's why it needs to be renewed every few years to respond to social demands.

4.1. Education

The concept of "education" is so old, and although its etymology (it comes from the Latin: *educatione,* which is the act or effect of educating, and from the verb *educare:* which is to develop the physical, intellectual and moral faculties) offers a clear understanding of what it is and what it is intended to be. Thus, education has always been and continues to be conceived as that which develops the individual's physical, intellectual and moral faculties.

In fact, in modern times, Rousseau (1712-1778) defined education as *a process of helping children to develop their strengths, so that they would be able to develop their abilities. He* also said that the child is born good but becomes corrupt as a result of the bad education given by a corrupt society (ROUSSEAU, 1995, pp. 11-13). These capacities can be physical, cognitive, affective-emotional, moral, among others, which lead to students' autonomy. To this end, the teaching-learning process and methodologies must always be innovative.

Etymologically, the term *education* derives from the Latin, *educatione,* from the verb *educare,* which means: to raise and nurture. And *ducere, which means* to lead. Education is a process of integrating an individual into a culture, (LIBÂNEO, 1998, p. 17). We realise that education is an intentional activity because it aims to promote learning. It takes place in space and time between the educator and the learner.

For Polak, Education, in its etymology, from *educare* (act of creating, nurturing) or *educere (to* lead out), indicates a very intimate and connected relationship between educator and learner, both influencing and transforming each other (POLAK, 2000, p.12). However, while education can take place systematically or unsystematically, teaching is a deliberate and organised action.

Therefore, it is systematisation that differentiates traditional education from systematic education with its own methods.

Luzuriaga, defining education in his work *"History of Education and Pedagogy"*, states that it is:

> the intentional and systematic influence on young people, with the aim of forming and developing them. But it also means the general, broad action of a society on young generations, with the aim of conserving and transmitting collective existence. Education is therefore an integral, essential part of the life of man and society, and has existed for as long as there have been human beings on earth (LUZURIAGA, 1978, p. 1-2).

In fact, education is such a fundamental component for every human being, because man is a social being. Since man is a social being, he needs the right training to ensure that he can integrate into the society in which he lives with others.

Education therefore aims to develop cognitive, affective, moral and social capacities so that people can be better integrated into society. It is through education that the ethical and moral values of a particular culture or society are preserved.

Rousseau, in his work *"Emilio ou de l'Education"*, strongly criticises the traditional education of his time, because it was too rationalised, too technical and too imposing. For these reasons, we can see why the philosopher Rousseau is considered the father of modern education - modern pedagogy, since he really takes the child seriously and recognises the importance of this stage. He starts from the principle that the child should be educated as a child and not as an adult.

In this sense, an education/pedagogy began that was child-centred and more democratic, no longer recognising the child as an adult, but rather their essence, their happiness and their freedom as a child. He determined that education should be provided from birth. Rousseau is considered the father of modern pedagogy, because he is one of the modern philosophers who brought innovative ideas, theorising a democratic education - centred on the student and following the respective phases he proposed. For this reason, education should be emphasised from the moment a child is born, as it moulds man and transforms society.

To better elucidate the definition, Rousseau shows the ancient meaning of the term:

> "Among the ancients, the word education had a different meaning from what we give it today: it meant nourishment. Thus, education, institution, instruction, are three things as different in their object as the ruler, the preceptor and the master. But these distinctions are misunderstood; and in order to be well orientated, the child must follow only one guide. We must therefore generalise our points of view and consider in our pupil the abstract man, the man exposed to all the accidents of human life (ROUSSEAU, 1995: 16).

In this quote, Rousseau says that modern man has distorted the true meaning of education, because among the ancients, the term education meant food, that is, nourishment. But the food referred to by the ancients was human values, scientific, social and cultural knowledge, among other things, passed down from generation to generation so that man could be satiated and behave in the best way within society.

Thus, education, the institution, instruction, are three elements as different in their object as the ruler, the preceptor and the master of the three. What is important, however, is that the child has only one guide, the one who accompanies the child in the process of teaching and learning. One of the principles of Rousseau's conception of education is freedom and spontaneity. Education should be carried out freely and spontaneously.

However, we can infer that education has taken on different meanings and significance from ancient times to the present day. However, it expresses a pedagogical doctrine that is based on the prevailing conceptions of man and society. It has emerged within and through the family, community, church, society and state.

Resseau says that the true meaning of education has been distorted. Among the ancients, the term education meant food. The food referred to by the ancients was, in fact, human values, scientific, social and cultural knowledge, etc., transmitted so that man could be satiated and behave in the best way within society. The educational thinkers mentioned here unanimously state that education is a process that takes place between educators and students in time and space. Both in its guiding ideas and in its actions, education has always interacted and interacts with the social, cultural, political, religious and economic transformations of each period of history and the development of humanity itself,

in its actions, it is influenced by these factors, just as, in a way, it also influences them. Far from having as its sole and main objective the individual and his interests, it is above all the means by which society perpetually renews the conditions of its own existence and, since society can only live if there is sufficient homogeneity among its members, education appears as the instrument that perpetuates and reinforces this homogeneity.

Therefore, for GOLDBERG (1998), *"to educate is to transform; it is to awaken aptitudes and orientate them for the best use within the society in which the learner lives;"* it is to develop cognitive structures that allow the individual not only to read and understand the world in which they live, but to act and, if possible, generate progress in society as a whole. This is the aim of any teaching or education, and maths education is not exempt from this.

This approach to the concept of education leads us to maths education, as these two expressions are interdependent.

4.2. Maths Education

In European countries, Maths Education, also known as Maths Didactics, is seen as an area of science dedicated to the study of the teaching and learning of maths.

In ancient times, there were concerns about the teaching of maths or mathematical education, particularly in Plato's *Republic VII,* but it is specifically in the Middle Ages, the Renaissance and the early Modern Age that these concerns are better focussed and developed. In particular, the focus is on Louis Antonio Verney in his 1746 work *Verdadeiro método de estudar (The True Method of Studying),* when he talks about the teaching of maths. However, it wasn't until the three great revolutions of modernity, such as the Industrial Revolution (1767), the American Revolution (1776) and the French Revolution (1789), that concerns about the mathematical education of young people or students began to take shape and be effectively addressed (MIGUEL, et al, 2004, p. 71).

Thus, as we described above, *the identification of maths education as a priority area in education occurred in the transition from the 19th to the 20th century* (MIGUEL, et al, 2004, p. 71).

The steps that opened up this new area of research are due to John Dewey (1859-1952, American philosopher), who proposed in 1895, in his book *Psychology of Number,* a reaction against formalism and a non-tense but co-operative relationship in teaching between student and teacher, and an integration between all disciplines, i.e. interdisciplinarity.

Furthermore, MIGUEL et al (idem, p. 72) argue that the most important step in establishing maths education as a discipline is due to the contribution of the eminent German mathematician Felix Klein (1849-1925), who published a seminal book in 1908, Elementary Mathematics from an Advanced Point of View. Klein, quoted by Miguel et al (ibdem), advocates a presentation in schools that is more psychological than systematic. He says that the teacher should, so to speak,

"to be a diplomat, taking into account the student's psychological process in order to capture their interest". He says that teachers will only be successful in their professional careers if they present things in an intuitively comprehensible way. Teachers should not see themselves as *magisters dicit,* i.e. holders of knowledge in the classroom. They have to deconstruct and then construct knowledge so that students can understand and assimilate the mathematical content.

Therefore, according to MIGUEL et al (1994, p. 72), the consolidation of maths education as a sub-area of maths and education, of an interdisciplinary nature, occurred with the foundation, during the International Congress of Mathematicians held in Rome in 1908, of the International Commission on

Mathematical Instruction, known by the acronyms IMUK/ICMI, under the leadership of Felix Klein. Thus, it was only after the creation of IMUK/ICMI at the 1908 International Congress of Mathematicians in Rome that the search for a suitable space for mathematical education in schools began to take shape. This is a giant leap forward in mathematical research, especially in maths education. This is the history of the development of maths education, from antiquity to the present day.

With this research, appropriate methods are proposed to respond to society's demand for quality education.

Maths is a complex subject that is difficult to assimilate and master for most secondary school students. This difficulty emerges from the very first steps of school life, and sometimes it is cultural, others it is acquired inadvertently at school, fundamentally in the initial classes of basic education, a fact that worsens over the course of their schooling. This research aims to summarise some of the ways of working in Maths Education that have emerged in recent times. Its purpose is, together with the bibliographical references, to contribute to those teachers who are concerned about Maths Education, but have not yet defined their working guidelines as a theoretical foundation for their educational activities at school.

Obviously, several thinkers have addressed the problems of learning maths in today's schools. From this perspective, TINOCO (1991) lists some of these problems, namely aversion and fear of failure.

These problems often lead the teacher to a situation of comfort in *"absolving him or herself of responsibility for the absence of learning and guaranteeing the attendance and discipline of students in their classes"* (ibdem). In other words, the teacher doesn't feel responsible for the students' lack of mastery and understanding of mathematical content. For the teacher, it's enough that the students are present and maintain discipline in the classroom. However, this teacher should be concerned about the lack of mastery and assimilation of maths and therefore look for other ways to teach his classes.

As a result of this comfort, in the author's view, the teacher remains forever in his or her (traditional) working method, indifferent to the difficulties presented by the students, since the teacher often shirks his or her responsibilities, thinking that there is nothing he or she can do to change the situation.

Another thinker who mentions some of the causes that jeopardise the teaching-learning process is Chagas. CHAGAS (s/d., p. 241) , states that there are many causes that contribute to the failure of maths education and more:

- Inadequacy of maths teaching in relation to the content, the working methodology and the environment in which the student in question is placed;
- The "bad" training of teachers, in other words, a lack of teacher training;
- Maths programmes are not flexible and are often based on models from other countries and, consequently, are models that often do not represent the socio-economic reality of the country;
- Lack of understanding and mastery of the fundamental prerequisites that would help this student develop well in maths lessons;
- Socio-economic devaluation of teachers.

From this perspective, the causes mentioned above have a negative influence on students' maths learning. Therefore, it is necessary to develop joint, innovative work, ongoing teacher training and appropriate methods so that the teaching-learning process takes place in a meaningful way.

According to CARRERA et al (1991), the traditional method has been the main methodology used in maths teaching and contributes negatively to the student's educational achievement. In fact, this method leads learning to a process in which *"some triumph and the vast majority fail"*. It's obvious that some students succeed and the vast majority fail, because the few who succeed are the most dynamic and understand the subject, while those who fail understand less mathematical content. However, this failure in maths is constantly seen in everyday school life.

Furthermore, IMENES (1989) points out that this fact is closely related to the formalisation of the teaching of this subject (mathematics), which has not changed over time, despite the transformations undergone by mathematical ideas in recent times.

In this sense, we have tried to develop some ways of finding solutions to the problems of Maths Education. Thus, we present some of these forms that are currently standing out in the various forms of teaching work, based on different theories or presented under different epistemological positions, with the incessant concern of improving mathematical learning in schools. These are: Critical Mathematics Education; Ethnomathematics; Modelling; Computers and Writing in Mathematics.

According to LOPES and BORBA (1994, p. 51) we see in *Critical Mathematics Education* the emergence of a constant concern to lead the pupil or student to question the society in which they live, i.e. the daily problems of the society in which they are inserted; in *Ethnomathematics* when knowledge "comes out" of the cultural context in which the pupil is inserted; meanwhile, in *Modelling* when one tries to write down in mathematical language a real problem in society; and in the use of

Computers when trying to bring modern technology to school that satisfies the anxiety for novelty that happens in the scientific world, or even when in *Writing in Maths* we see the concern to give everyone opportunities to express their thoughts, reflecting and expressing their own languages and opinions.

In this context, we have tried to briefly describe each of the aspects of Maths Education mentioned above.

4.3. Critical Maths Education

In the view of Lopes and Borba (1994), the school needs to fulfil a very important function rather than just preparing the pupil or student for work or the job market. Therefore, the school needs to prepare the pupil or student for life in society; it needs to teach how to think and educate how to act, in other words, it *needs to educate students to become critical citizens of the social structure of which they are a part* (ibdem).

Educating pupils or students to become critical and/or autonomous citizens of the social structure is the task of education as a whole and maths education in particular. To this end, it is necessary for this education to be democratic and liberal, where the diverse opinions of the students are accepted and debated together. The classroom should be conceived as a laboratory of knowledge, where each student finds a space to express their thoughts and then have them analysed.

It's from here that the student gains the strength to speak in public, learns to organise their thoughts and to respect other people's thinking, because no one is the master of knowledge, both the teacher and the students, everyone is trying to learn more and more. That's why everyone learns in the classroom.

LOPES & BORBA (1994, pp. 51-52) state that Critical Maths Education, whose main ideas are based on Critical Education, is a way of working with maths that directs educational practice and research towards trying to overcome social differences. It endeavours to direct the educational process towards the pursuit of *"critical competence"*.

This *"critical competence"* would give the pupil or student the ability to participate in the democratisation of society, to give their opinion on a given public issue and to construct their own knowledge. However, critical maths education would form a critical and autonomous student.

We have seen in schools that the majority of students taking maths or maths courses ask themselves questions like: what is this for? Where am I going to apply that? These questions are not answered because the mathematical content is perceived to be disconnected from the student's daily reality. As a result, this leads to students becoming complacent. With the traditional method, the teacher prepares his or her lesson, lectures, explains and draws up diagrams. While the student memorises everything the teacher says.

In fact, to further substantiate the issue, we can see that in schools where maths teachers work with traditional teaching, the students' teaching-learning process becomes mere transmission of the subject, i.e. the teacher *transmits* and the students *receive*. However, this activity of transmission and reception is accompanied by the repetitive and purely mechanised performance of exercises, resulting in the student memorising how these exercises were initially developed and inhibiting the development of their cognitive faculties, abilities and skills. And this weakens the student's learning.

This practice deprives or underestimates the student's cognitive capacity. This is why critical mathematics education, at a time when the teaching-learning process is mechanically memorised, is necessary in order to train critical and autonomous students who are, above all, capable of constructing their own knowledge. Therefore, theoretical advances show that the teaching-learning process does not take place through decontextualised mechanical training or teacher exposure. On the contrary, concepts are learnt through students' interaction with knowledge. We are therefore talking about education centred on the student and not the teacher. The student is the protagonist of their education and the teacher can simply guide them through this educational process.

It is essential to note that the teaching-learning process is made up of various activities or factors that must be organised in advance by the teacher, with a view to the assimilation of knowledge, skills and habits by the students, the development of their intellectual capacities, always aiming for mastery of knowledge and skills and their various applications (CHAGAS, s/d, p. 246).

For Chagas, the key within the teaching-learning process is to change how we *teach to how students learn and what I do to favour this learning* (ibdem). To do this, we must realise that mathematical content guides the teaching-learning process, prioritising individual (student) and collective (society) construction. We therefore need to create situations in which students interact with the object of knowledge and establish their hypotheses so that they can then be confirmed or reformulated by their classmates and the teacher.

Chagas goes on to say that:

> I believe that the first step to be taken is for maths education to break away from the traditional model and opt for a more constructivist context, where students have to analyse a given problem before they can understand it. It is important here that the teacher offers space for discussions and continually interacts with their students (ibdem).

This passage makes it clear that the first step towards success in maths teaching is to completely separate maths education from the traditional model, increasing constructivism in the teaching and learning process.

In this context, the student will be able to analyse a particular object of study or situation, and then the students will be able to understand it. To do this, the teacher needs to provide spaces for discussions and constantly interact with their students. In addition, for teachers to be able to stimulate their pupils, they need continuous training beyond the relevant basic professional training they have; teachers need to realise that pupils must feel interested in solving problems in order to learn mathematics, as this awakens the pupil's curiosity and creativity.

Therefore, we infer that the main objective of Critical Mathematics Education is to get pupils or students to hermeneuticise reality in such a way that they are able to organise themselves to intervene in the social and political context in which they are involved, in other words, they should be citizens who actively participate in solving the problems that afflict society.

SILVEIRA (2006, p. 4) in his article *"Criticism of Maths Teaching"* argues that nowadays, due to the inference of new theories such as constructivism, the maths teacher is required to illustrate to the student how mathematical content or mathematical subjects can be related to the student's daily life. However, this isn't always possible due to the teacher who lectures and doesn't teach, and the student who doesn't care or doesn't make the effort to assimilate the mathematical content.

In this context, it is necessary for the mathematics teacher to make or undertake a great effort to achieve this goal, and *the student must strive to believe that everything around them can be mathematised* (SILVEIRA, 2006, p. 4). This means that students don't have to think that mathematical content can't be assimilated. They can be assimilated if they make the effort to do so.

4.4. Ethnomathematics

We start from the principle that if maths can be found outside of school, then so-called *ethnomathematics* can be used effectively in school as a way of exploring our students' knowledge, leading them to the

academic knowledge (LOPES & BORBA, 1994, p. 55). In other words, the teacher uses the students' non-scientific knowledge as a starting point for scientific knowledge. But it's important to emphasise that for this to happen, they must start from the context to which the student belongs, not using the reality of a group unknown to them, as this will make it even more difficult for them to understand and assimilate the mathematical subject.

In fact, BORBA (1987) illustrated this through his experience in a neighbourhood in Campinas (Brazil). This thinker observed that the children, even though they had no contact with the school, were carriers of mathematical knowledge and this knowledge was not as far removed from scientific maths as one might assume from a superficial, in-depth observation. For this reason, this knowledge is important and the classroom teacher should value it, considering it as the starting point for their lesson.

Therefore, the search for maths in the students' culture, in their daily lives, present in ethnomathematics, can be an important strategy for teachers, not only as a captivating element but also as a teaching methodology in maths and perhaps in other subjects.

Ethnomathematics relates to mathematical modelling in terms of the tactics it uses as teaching strategies in the educational process.

Berry defines mathematical modelling as: *"[...] the process of choosing characteristics that adequately describe a problem of non-mathematical origin, in order to put it into mathematical language"* (BERRY *apud* LOPES & BORBA, 1994, p. 55).

In fact, mathematical modelling is another way of trying to understand mathematics in students' everyday lives,

because it is an attempt to translate a problem that has arisen in the real world into mathematical language - as a way of solving it as precisely as possible. In this context, the teacher takes the problem from the student's everyday reality into mathematical language. So these are some ways that can boost maths education if they are well applied in the teaching-learning process.

4.5.   Mathematical modelling

Mathematical modelling is a simple way of getting students to reflect on everyday realities using mathematical knowledge. These investigated extra-mathematical realities are then expressed in mathematical language.

In this way, students learn to make connections between the realities of everyday life and school maths.

Berry defines mathematical modelling as: *"[...] the process of choosing characteristics that adequately describe a problem of non-mathematical origin, in order to put it into mathematical language"* (BERRY *apud* LOPES & BORBA, 1994, p. 55).

In fact, Mathematical Modelling is another way of trying to understand mathematics in the student's everyday life, as it is an attempt to translate a problem that has arisen in the real world into mathematical language - as a way of solving it as precisely as possible. In this context, the teacher takes the problem from the student's everyday reality into mathematical language. So these are some ways that can boost maths education if they are well applied in the teaching-learning process.

Looking at the historical context of mathematical modelling, we find, in ancient times, some researchers who used this trend (mathematical modelling) to arrive at an expected result.

One of the scholars we can cite is Thales of Miletus (naturalist philosopher, 540 BC).

As stated in the writings of Thales of Miletus, there's no need to climb the pyramid of *Cheops to* calculate its height, you just need to stick a stick in the ground and use the idea that its shadow will be projected onto the ground, forming a

triangle, which is similar to the face of the pyramid, making it possible to calculate its height. By making a hermeneutic of the above Thalesian thought, we can infer, through this measurement methodology, that Thales empirically practised what is now known as Modelling.

According to the example cited, we can see that modelling has been applied since ancient times. And even today, some schools apply modelling in the teaching-learning process, particularly in mathematics.

Therefore, modelling is one of the trends in mathematics teaching that can contribute to the construction of knowledge; as a differentiated method that students can use in the teaching process to interact with the environment in which they are inserted, thus preparing them as critical, participative and reflective citizens. Modelling makes it possible for maths teaching to have contextualised themes - which start from the student's reality, because it awakens in the student reflection and a critical sense as an active and autonomous being in the society in which they live.

Currently, some considerations are made by scholars, including BASSANEZZI (2004, p. 16) who states the following:

> Mathematical modelling in education is the art of expressing, formulating, resolving, elaborating expressions through mathematical language, everyday situations and later serving other areas. Various fields use modelling for their research, including: Physics Chemistry, Biology, Astrophysics, among others.

The quote makes it clear that mathematical modelling is conceived as the art of expressing reality, formulating hypotheses based on the expression of reality, solving everyday and mathematical problems and, finally, developing expressions using mathematical language based on everyday situations. Modelling is not restricted simply to the discipline of mathematics, but is also used in the fields of physics, chemistry, biology, technology and other fields of knowledge.

According to BARBOSA (2007, p.161), mathematical modelling is a teaching and learning environment, *"[...] in which students are invited to investigate situations with reference to reality through mathematics"*.

Based on Barbosa's ideas, we can consider that mathematical modelling allows for more active student participation, the possibility of confronting opinions and stimulating the presentation of alternatives to solve problems, thus developing cognitive abilities that allow them to understand, analyse, elaborate and create new situations aimed at forming a critical citizen capable of reading the reality that surrounds them, from the different political, economic, historical and cultural dimensions with a view to social transformation.

In this respect, mathematical modelling appears, in the teaching-learning process of general secondary education, as a different way of teaching that can allow students to be agents in the construction of their own knowledge, overcoming the difficulties that mathematics presents in understanding content, its concept and its representation through graphics, based on a reality experienced in everyday life, because the student will be learning meaningfully. On the subject of meaningful learning, AUSUBEL, NOVAK and HANESIAN (1978, p. 90) state that:

> The essence of meaningful learning lies in the fact that ideas expressed symbolically are related to information previously acquired by the students through a non-arbitrary and substantive relationship.

However, meaningful learning takes place when the teacher considers the student's prior knowledge about the content taught in the classroom to be important, because the teacher and the students link prior knowledge with scientific knowledge. In this way, students learn not in a mechanical way, but by relating the content learnt in the classroom to their everyday lives. Consequently, this helps students to better understand the importance of what they are learning by relating it to the reality of everyday life.

According to the Vygostian perspective, we can say that learning takes place in the joint activity of the teacher and the students through activities that include characteristics of the content, the specific task and the quality of the students' interaction.

In a country like Mozambique, the main challenge is teacher training and practice, so that methodological alternatives are recognised and adopted that change or improve the relationship between students and mathematics at school, enabling them to learn. With regard to this perspective, BASSANEZI (2013, p. 43) states that:

> The biggest difficulty we've noticed in getting most maths teachers to adopt the modelling process is overcoming the barrier naturally created by traditional teaching, where the object of study is almost always well-defined, following a sequence of prerequisites and with a clear horizon for arrival, which is often compliance with the subject's syllabus.

Modelling is an innovative proposal. It is a teaching and learning strategy that allows the teacher to engage in interesting dialogue with the students when explaining the content, in order to provide tools capable of leading students to form their own concepts and solve problems using models that facilitate their learning.

Various authors refer to the use of this teaching and learning methodology in mathematics as a differentiated alternative to the traditional method - where the teacher's posture continues to be the centre of the lesson, with the sole function of transmitting pre-defined content with an outlined and finished objective, making the student passive in the teaching and learning process. In this sense, researchers defend the paradigm shift as a way of providing students and teachers with a new vision of understanding how learning processes occur, i.e. a way that helps the teacher transform information into knowledge as a mediator and facilitator of the teaching process.

Likewise, the same authors discuss mathematical modelling as a motivator in teaching practices in order to develop reflective thinking and meaningful learning in students and make them critical, reflective, participative citizens capable of defending their own ideas in the contemporary and competitive world, where, according to Bassanezi, one of the pioneers in research into mathematical modelling in teaching, specifically in Brazil,

"Mathematical modelling is the art of transforming problems from reality into mathematical problems and solving them by interpreting their solutions in the language of the real world" (BASSANEZI, 2006, p.16).

Bassanezi sees mathematical modelling as a simple way of transforming everyday problems into mathematical problems and trying to solve them using mathematical language. This is an activity that arises from the students' initiative to investigate a particular extra-mathematical problem, but as they solve it, they explain it using mathematical language to express the results obtained from studying the problem.

In addition, other researchers, such as Barbosa, present more specific definitions of mathematical modelling with a focus on Mathematics Education. He conceptualises mathematical modelling as *"a learning environment in which students are invited to investigate, through mathematics, situations with reference to reality"* (BARBOSA, 2007, p.161).

In this sense, according to the authors mentioned above, we infer that modelling proposes effective changes to traditional teaching practices, opening up a way to awaken students' interest in the content dealt with in the classroom, relating the problematic circumstances of everyday life to the construction of possible solutions, using mathematical concepts.

In fact, Mathematical Modelling has been referred to by various mathematical educators as a pedagogical alternative that aims to relate school mathematics to non-mathematical issues of interest to students, configuring an activity that is developed according to a scheme, in which the choice of the problem to be investigated has the direct participation of the subjects involved - groups of students. These groups of students choose, under the guidance of the teacher or not, a theme or object as a problem to be solved, define the objectives, present the justification, formulate possible hypotheses, research and then present the results of the research expressed in mathematical language.

5.5.1 Constructivism as a perspective on teaching and learning

This is the model most frequently addressed in research on science teaching in the early years of primary school, predominating in more than half of them according to Fahl (2003) and Fernandes and Megid Neto (2012). It is a view shared by different

This is a model to be followed on the fronts of science teaching research (ASOKO et al., 1999; KRASILCHIK, 1992; BORGES, 2002). It emerged in the 1970s from a rich context of "bringing pedagogy closer to epistemology, sociology, psycholinguistics and the history of science" (FAHL, 2003, p. 46), and also after realising that the rediscovery model did not bring significant advances in relation to the traditional model.

In addition to the transformations in pedagogy itself, the unquestionable character of science and its role as a social agent and dominator is more debated within sociology and epistemology (LATOUR; WOOLGAR, 1997; BOURDIEU, 2004), and this discussion is incorporated into the model (KRASILCHIK, 1987). The emergence of the cognitivist perspective in pedagogy is also a pedagogical trend that guides this model; the model is therefore supported by classic authors such as Piaget, Ausubel and Vygotsky (FERNANDES, 2015). There are some variations within this model, not necessarily confluent, but the construction of knowledge by the student is central, as is the mobilisation of previous conceptions (ASOKO et al., 1999). The student's previous and cultural knowledge is used in the construction of knowledge and is the target of confrontation (not necessarily litigation) with scientific knowledge. From Libâneo's (2009) perspective, it is close to progressive trends, and also to the cognitivist approach, as indicated by Mizukami (1986). Fernandes (2015) brings the idea of continuous construction of knowledge into the model:

In the constructivist model, knowledge is considered to be a continuous construction and the transition from one stage of an individual's development to another is always characterised by the formation of new intellectual and cognitive structures that did not previously exist in the individual. (FERNANDES, 2015, p. 121)

This model works with the concept of science and the nature of scientific knowledge, articulating these notions in the construction of knowledge. Science is not presented as the source of truth.

In addition, the model conceives of knowledge as something social and cultural, so these two aspects need to be taken into account in the learning process, and the history of scientific knowledge is contextualised (KRASILCHIK, 24, 1987). The student is seen as an active and critical subject, so their mobilisation is fundamental, while the teacher is a guide to the process, not the holder of the truth. The teacher plays a fundamental role in this vision, but not as an indoctrinator of behaviour, as he or she is a mediator of discovery and not just a transmitter of content, using argumentation and highly conscious preparation of lessons (BORGES, 2002). The teacher's argumentation is paramount, as the student is challenged all the time to reflect, structure hypotheses and solve problems; this is coupled with the strong social interaction between students, which also incorporates the interactionist current into the model (FERNANDES: 2015).

5.5.2 Mathematical modelling as a way of changing the teaching-learning process

As we said earlier, from ancient times to the present day, maths books have included certain problems for readers to solve. Archaic books or texts, such as those of the Egyptians, Babylonians and Chinese, were already composed of a list of problems whose solutions were subsequently provided (after research or investigation). In this context, on the one hand, the problems were chosen as a way of teaching the reader about maths, and were often posed by degree of difficulty; on the other hand, these problems generally reflected the needs of societies, different aspects of daily life or problems that troubled society (VIECILI, 2006, p. 25).

Therefore, it is understood that books with mathematical problems have appeared in all civilisations throughout history right up to the present day. However, what concerns us is the fact that we see the same problem appearing in texts from different civilisations and at different times in human history. And sometimes these problems appear decontextualised, artificial, repetitive and with distorted content. To deal with this scenario, teachers need to be trained regularly so that they are equipped to work with mathematical modelling, contextualising the content so that learning is meaningful.

Thus, when working with mathematical modelling in mathematics teaching, two aspects are totally important:

i.   Combining the chosen topic with the students' everyday reality; and

ii.   Making the most of out-of-school experiences, linking them to classroom experiences (ibdem).

These two aspects are fundamental and interconnected. One should not work without the other. The teacher dealing with mathematical modelling in mathematics teaching must follow them scrupulously.

For PINKER, quoted by SCHEFFER (1995), mathematical modelling has the following phases:

1^a : Formulating the problem;

2^a : Building a mathematical model;

3a Searching for and testing a model solution e,

4a: Finally, validation of the solution.

However, the author goes on to say that information, questions and assessment criteria are prerequisites that the teacher must use when teaching mathematics to construct a modelling problem.

Furthermore, there are consensus writings that argue that mathematical modelling brings various benefits to the teaching-learning process. GAZETTA (1989), one of those who advocates the benefits of mathematical modelling, lists the following:

i.   Motivation on the part of the student and educator;

ii.   Ease of learning - mathematical content goes from abstract to concrete;

iii.   Due to the interactivity of content, preparation for future professions in the most diverse areas of knowledge;

iv.   Development of logical thinking;

v.   It gives the student the opportunity to be a critical citizen and transformer of their reality; e,

vi.   Understanding the socio-cultural role of maths, thus making it more important.

Based on these general concepts, the intention is to illustrate the importance of maths for human knowledge and understanding the reality in which individuals find themselves. This does not mean that Mathematical Modelling should be used as the sole and exclusive teaching methodology at school. Other methodologies can be adopted, such as constructivism (the student produces knowledge), as this has "almost" the same characteristics as mathematical modelling.

Teachers, when carrying out their teaching activities, should always look for the best teaching methodology, involving games, play, in short, using all their resources to get the best possible result when teaching maths. Nevertheless, the arguments presented should not be disputed, as they are in favour of Mathematical Modelling, as can be seen below:

i. Motivation for the educator and the student;

ii. Facilitating learning;

iii. Preparation for using maths in different areas;

iv. Skills development; e,

v. Understanding the socio-cultural role of maths.

In parallel with the above, BARBOSA, quoting BIEMGENTUT (1990), argues that modelling activities can be seen as a way of mathematically educating students to exercise the rights of citizenship, thus challenging the ideology of certainty and placing critical ideas on the applications of mathematics - critical and reflective knowledge.

It is clear, however, that the field of modelling is associated with problematisation (questioning or problematising something) and research (searching for, selecting objects or information for study, organising and manipulating information). Thus, in this field, as well as applying the knowledge that has already been built up to real situations, there is the possibility of acquiring new knowledge while working on mathematical modelling or in another area of knowledge.

According to SCHEFFER (1995), *"Modelling is a learning environment in which students are invited to problematise and investigate, through mathematics, situations with reference to reality"*.

Herein lies today's "Achilles heel" in the teaching-learning process, in other words, one of the great challenges of our times, which is to make students understand the importance of their role in society, as active, participative agents who transform their surroundings, and the relevance of maths in their daily lives.

Roughly speaking, in the author's view, there are a number of challenges to be overcome in the teaching-learning process:

- The lack of support from educational institutions to provide the necessary and sufficient conditions for new practices;
- Demotivation on the part of teachers who work too many hours;
- Students' lack of interest and indiscipline;

- The lack of time to develop alternative teaching projects.

According to ALMEIDA (1993), there is also resistance from some maths teachers who are "used to" traditional teaching and are opposed to trying new methodologies and the possibility of changing their practices. In this sense, such resistance prevents, for example, the exploitation of resources offered by Mathematical Modelling and Constructivism in teaching practice with mathematics, making it more meaningful in that it would be valuing both everyday situations, or situations related to other areas of knowledge, and the student's understanding.

What's more, the syllabus is pre-established and often doesn't give teachers the opportunity to vary their teaching methodology because, from their point of view, they have to "comply" with the syllabus - which is inflexible and not dynamic.

From the perspective of BASSANEZI (2002), Mathematical Modelling can be considered as one of the pedagogical paths that arouses interest, expands pupils' or students' knowledge and helps them to structure the way in which they think, reason and act towards the object or information of knowledge.

However, when it comes to mathematical modelling from an epistemological perspective, ALMEIDA et al (2011) argues that:

> [...] Mathematical modelling, besides being mathematics, is also epistemology, since mathematical models aim to understand and explain facts and phenomena observed in reality, as quoted by Bassanezi, i.e. knowledge and understanding of that reality (CIFUENTES; NEGRELLI, 2009, p.45).

According to the excerpt, the questions that arise for both teachers and students are the following: how does the interaction between the object (the chosen theme) and the subject (the group of students) take place when working with modelling? And how does scientific knowledge come about in modelling activities?

What ways are used to solve problems with modelling and what novelty do they bring? These questions serve as leverage when carrying out a modelling activity. With these questions, modelling can be related to constructivism, as both advocate that knowledge is constructed by the subject (student). However, students should not passively receive knowledge. They have to criticise it in order to become autonomous, reflective, active and participative both in the classroom and in the democratisation of society.

In fact, the student should not be a "copyist" of the content covered, but should build their knowledge on it. This is why modelling and constructivism come together, as they can be used as methodologies in the teaching and learning process.

Moreover, referring to the importance of constructivism in school mathematical activities with Modelling, ALMEIDA and VERTUAN (2011, pp. 25-26) state that:

> By making use of maths, considering both the use of algorithms and mathematical concepts themselves, students can apply knowledge they have already built up during lessons, or construct new knowledge. In many situations, when involved in modelling activities, students come up against an obstacle for which they temporarily do not have enough knowledge to overcome it, and so the need to construct this knowledge through this activity emerges. Therefore, in modelling activities, students can both re-signify concepts they have already constructed and construct new ones when they need to use them.

This means that students can construct their knowledge using mathematical language, but in order to do so, it is essential that they have first practised the application of constructivism and modelling in the subject of mathematics. That's why we say that these two terms (constructivism and modelling) are related in their application. Otherwise, as they go along, they will come up against barriers and may not be able to realise their activity. Therefore, the teacher plays an important role in the teaching-learning process in terms of explaining the applicability and functionality of these methodologies in the classroom, as he or she appears as a mediator and guider of the process.

According to SILVEIRA (2005, p.71), *"[...] teachers are constantly experimenting with new techniques that make it possible to develop students' abilities to acquire mathematical knowledge"*. Silveira also argues that "in addition to constructivism, there is the possibility of working with mathematical modelling", which according to BEAN (2001, *apud* SILVEIRA, 2005), *"has an approach in which students create mathematical models to represent given situations. In other words, it is similar to problem-solving procedures"*.

With this in mind, the author clarifies the following: on the one hand, there are teachers who develop by experimenting with new techniques or methods in order to develop students' abilities to obtain mathematical knowledge. However, on the other hand, there are teachers who resist applying new teaching techniques or methods, claiming that such methods (constructivism and modelling) take too long and that they have to comply with the syllabus. In addition, there are students who are uninterested in learning.

According to BARBOSA (2007), in a learning environment from the perspective of Mathematical Modelling, the teacher must be a questioner and the one who teaches to question. In this sense, Barbosa's conception is based on the understanding of Freire and Faundez (1998), according to whom:

> What teachers should teach - because they should know this themselves - is, first and foremost, to teach how to ask questions. Because the beginning of knowledge, I repeat, is to ask. And it is only from asking that one should set out in search of answers, and

not the other way round (FREIRE; FAUNDEZ, 1998, p. 48).

Therefore, from the author's perspective, the aforementioned thinkers present constructivist conceptions, whereby through modelling activities, students not only build knowledge of other areas of the school curriculum, but are also able to formulate their own research questions in order to seek answers to the problem situation being investigated or questioned. It is extremely important at this point to explore the importance of mathematical concepts.

At this point, students practice mathematising everyday realities with the mathematical concepts used and assimilated in the classroom. It is therefore important for teachers to prepare students for the use of mathematical constructivism and modelling, as this will enable them to develop their skills and abilities, moving from theory to practice, transforming society and producing new knowledge as a way of actively participating in the construction and transformation of the society in which they live. This reflection leads us to the constructivist conception of the teaching and learning process in the classroom.

5. Articulation between modelling, constructivism and the curriculum plan for the general secondary education of Mozambique

The idea that students are active subjects in their learning and development process is based on the idea that they are capable of thinking about the situations presented to them and formulating hypotheses (often erroneous), reasoning and establishing relationships based on their experience, their own ideas and intuitive knowledge. Modelling, which in turn requires students to think logically and test solutions to problems, shows the possibilities of mobilising students' interest so that, while formulating and testing solutions to one or more problems, they can establish a relationship between mathematical notions and a set of everyday situations. There is a common denominator between Constructivism and Mathematical Modelling that makes it clear that there are links between two processes identified with the student's cognitive experience, with their understanding and with meaningful learning. This suggests the possibility of an approach to teaching mathematics based on bringing together the assumptions of mathematical modelling and Constructivism.

constructivism, allowing us to suggest mathematical *modelling in a constructivist approach* as a guideline for teaching *mathematics* in basic education.

This articulation is supported, in our opinion, by Biembengut's (2016) mapping of Mathematical Modelling productions in Brazil, which resulted in the grouping of three conceptions of Modelling in Education, namely: 1) method or strategy (realistic and epistemological), 2) pedagogical alternative (contextual and educational) and 3) learning environment (sociocritical). For the author, these conceptions "represent the sum of contributions from many teachers interested in improving school learning, enhancing knowledge and better living and acting in society." (Biembengut, 2016: 169).

According to the author:

The concept of modelling adopted by the authors in their teaching experiences has a common precept: to make students more interested in maths lessons based on what they understand, experience and can participate in, whether based on their previous knowledge or their beliefs.

And the trends identified suggest that in classroom practices the proposals have sought to encourage students to become actively involved in their learning, produce work from personal needs, interests and goals in a challenging and talented way and lead to rich humanitarian commitments (BIEMBENGUT, 2016, P. 170).

When the general objectives of the curriculum plan for general secondary education in

Mozambique are analysed, it can be seen that there are approximations and convergences between them and what was previously highlighted as an element of articulation between constructivism and mathematical modelling.

Changing pedagogical practices in the classroom simultaneously requires changing paradigms or methodologies in the teaching-learning process of maths in accordance with the general objectives outlined in the curriculum plan for general secondary education in Mozambique (INDE/MINED, 2003) [...],

because they point out that the student should be able to:

- Recognise mathematical knowledge as a means of understanding the world around them through research and the development of actions that stimulate interest, curiosity and the resolution of everyday problems;
- Recognise that mathematics is a very useful tool for life and is an integral part of our cultural roots because it helps us to think and reason correctly;
- Develop the ability to communicate with others;
- Solve mathematical problems that reflect everyday situations in the economic and social life of the country and the world;
- Developing skills for searching for information in different media and using technology, showing curiosity and a willingness to seek out new knowledge;
- Developing self-confidence, expressing and arguing their opinions, making judgements about concrete situations, facing new situations with confidence and showing flexibility and creativity in mathematical learning and beyond.

According to the objectives defined by the PCESG in the subject of Mathematics, it can be seen that the elements of an integrative approach to modelling and constructivism are tacitly included in the Mozambican school reality, as it is argued that students should be able to produce autonomous knowledge and relate mathematical knowledge to the reality of daily life. However, *modelling in the constructivist approach,* as a methodological proposal, can contribute to changing the teaching-learning process of mathematics in the classroom, through problem solving, which is increasingly recognised as a practice favourable to the development of competences. Because today it is inconceivable that human beings should be mere reproducers, not forgetting that with traditional teaching, students don't learn, but with the new trends of globalisation there is a need for them to be participative agents, with critical and creative thinking, equipped with the skills and competences to face up to the current reality of the contemporary world and stand up to it, because teaching is linked to the world of competition.

Given this scenario, the author proposes the use of modelling in a constructivist approach to the teaching-learning process.

Several studies (BIE-UNESCO) point to resistance to change on the part of the various players as one of the main obstacles to implementing a curriculum. Indeed, changing pedagogical practices in the classroom implies a change in attitude and positioning (in relation to the student as the centre of the teaching and learning process).

Resistance to change in education is generally motivated by various factors, including the following:

-Successful implementation is possible if you bear in mind that there are different phases that characterise the process of change:

a) Reaction against the need for change that breaks or alters habitual practices, aggressiveness/passivity;

b) A period of reflection and questioning of habitual practices;

c) Developing a critical spirit in relation to their usual practice / questioning practices (developing awareness of the need for change);

d) Personal responsibility for change and Sharing in the process of change (positive attitude towards change) ;

e) Internalisation of change; fear of making mistakes on the part of decision-makers. lack of specific training for those responsible fear of change lack of awareness of their role as a player in change; lack of communication between management and teachers complexity of innovations rigidity of structures and guidelines lack of coordination between the different sectors;

f) Fear of losing power because of the entry of other actors or a change in status;

g) Misinterpretation of official texts

h) Administrative problems;

i) Resistance to change, Beginning of the change process.

Changes at school are more difficult because we have internalised traditional teaching practices for many years.

For there to be a positive evolution in the process of changing pedagogical practices in schools, strategies need to be developed that promote moments of reflection by the teachers themselves on their pedagogical practices in the classroom. These strategies should be developed at school, district, provincial and central level, through coordinated actions. Changing behaviours, mentalities and traditional teaching practices requires a capacity for self-knowledge and knowledge of others. Self-knowledge is essential for effective communication in the process of change.

In this context, a teacher who is aware of how they perceive themselves and others will easily identify and internalise the process of change. For teachers to be able to develop competences in their students, they need to know themselves very well, to know their qualities, skills, defects and limitations in order to get the most out of their relationship with others and thus optimise their professional performance. This curricular transformation involves, in addition to the introduction of new curricular documents and new approaches, the possibility of changing established beliefs or practices. Therefore, the implementation strategy aims to anticipate and predict possible paths to the successful introduction of the curriculum and is an important tool in managing the conflict generated by the changes to be made. The curriculum implementation strategy is based on the following vectors:

1. Teaching-learning methodologies;
2. Teaching and learning conditions;
3. Involvement of the community and other education partners;
4. Teacher training.

- Teaching-Learning Methodologies

The teaching-learning methodologies in this study are based on the principles that guide the current curriculum for Mozambique's Basic School, namely:

a) Inclusive education;
b) student-centred teaching and learning;
c) learning orientated towards the development of life skills;
d) Integrated teaching and learning;
e) Spiral teaching and learning.

In the light of these principles, the teaching/learning process is organised taking into account that the student is an active subject capable of constructing their own learning. In this sense, students should have the opportunity to acquire and experiment with a set of tools that will enable them to develop their own world view and apply what they learn in life situations, both foreseen and unforeseen.

Students are not a homogeneous mass; each student is seen as unique, with varying learning styles and rhythms.

Therefore, the teaching and learning strategies to be adopted should be diversified and adjusted to the real needs of the learners. Special help for students is needed so that they can, on the one hand, develop study methods suited to their learning style and, on the other, work in pairs and groups to establish a dialogue between them to solve real non-mathematical problems. The teacher, as mediator, must create opportunities for students to develop the competences defined. Group work is seen as one of the effective strategies in student-centred teaching methods, as it contributes, among other things, to the development of social skills and increases levels of understanding and self-confidence since the students plan and manage the tasks between themselves.

In this sense, they will be encouraged to get involved in common projects at school and class level, as a way of managing the tasks between them. In this sense, they will be encouraged to get involved in common projects at school and class level, as a way of realising the interdisciplinary, student-centred approach aimed at developing life skills.

In the joint work between teachers, strategies for dealing with cross-cutting themes, teaching and learning values and developing competences should be improved, so that everyone is committed, takes on and engages in the education of young people.

The teaching-learning strategies also include preparing students and teachers to methodically search for information in libraries and using the new Information and Communication Technologies. Piaget developed a very complex theory in his quest to explain the genesis of knowledge. For Piaget (1987), all individuals, regardless of culture and social status, can go through the same development process, which takes place in four stages: sensory motor, pre-operational, concrete operational and formal operations.

But this doesn't happen in a linear order, as each stage begins by reorganising the previous ones, and this process can be slowed down, or even blocked, depending on the interactions established.

The stages of knowledge development provide indicators for defining the complexity of the situation, in other words, they provide favourable learning conditions for the student's current stage of development. Thus, for PIAGET (apud LA TAILLE; OLIVEIRA; DANTAS, 1992), learning is acting on the object of learning in order to understand and modify it. As learning is a continuous adaptation to the external environment, one learns when one enters into cognitive conflict, i.e. when there is a confrontation of ideas and situations of something that one does not know the best solution or answer to.

The individual endeavours to find equilibrium. For Piaget (87), the process by which schemas are changed-adaptation-is made up of two complementary aspects, assimilation and accommodation. From this perspective, the individual needs to go through an important process of adaptation, in which a balance is established between assimilation and accommodation.

Assimilation refers to the subject's action on the physical and/or social environment and therefore depends on personal initiative for the construction of knowledge. Through incorporation, the existing knowledge structure is modified in order to accommodate new elements - this modification is called accommodation. Equilibrium is the process of organising cognitive structures into a coherent, interdependent system that enables the individual to adapt to reality. It is from this understanding that learning situations are based on games and challenges, in which the subject is faced with a new problem to solve.

According to Blumer (1980), the concept of meaning from a constructivist perspective results from communication and the interaction of individuals with objects in the outside world, with other individuals and with themselves.

According to Blumer (1980: 22), the process of human interpretation has two distinct phases:

a) In the first, the agent (teacher) determines for himself the elements with which he relates, the individual establishes a communicative process internally; and

b) In the second phase, as a reflection of self-communication, interpretation is a mental process triggered by the student, aimed at understanding the content taught by the teacher.

According to Coll (1988), learning from a constructivist perspective presupposes a personal process of building knowledge, where the student's actions, experiences and own objectives allow them to interact with the physical and social environment, decisively conditioning new learning.

Coll's position shows that the acquisition of new knowledge results from the student's individual activity, in which the knowledge they have already acquired allows them to interact with what is taught or made available by the environment around them.

Therefore, in the process of teaching and learning maths, it is essential to take into account the experiences, ideas and explanations about natural and everyday phenomena that students bring into the classroom. Generally, these differ from scientific concepts, even though they are consistent, coherent and functional for a certain community. These ideas and experiences are called student conceptions.

According to Wandersee (1985), students' conceptions are brought into formal maths instruction:

a) They originate from a diverse set of individual knowledge or experiences about the nature of objects and events, acquired in the out-of-school environment through direct observation or perception, culture or language, as well as from the explanations of teachers or instructional materials;

b) They depend on age, gender, cultural boundaries, social or academic level;

c) They are resistant to extinction through traditional or conventional teaching strategies;

d) They offer parallel explanations to natural phenomena;

e) They are usually subscribed to by teachers; and

f) They interact with the new knowledge, resulting in new, unintended learning.

To remedy the student's conceptions, within constructivism based on conceptual change, new teaching-learning methods were born, including those based on: (i) cognitive conflict; (ii) analogies; and (iii) mathematical modelling. The teaching-learning method based on cognitive conflict consists of demonstrating differences between the student's conceptions and scientific conceptions, a process in which the student becomes dissatisfied with their own conceptions (Novick & Nussbaum, 1982).

The teaching-learning method based on analogies consists of creating bridges of transition between the student's conceptions and scientific conceptions (Gunstone & Mitchell, 1998).

The teaching-learning method based on mathematical modelling consists of manipulating objects from the environment or dynamic models on the computer to solve problems (Biembengut & Hein, 2003).

The latest research from the last century shows that the teaching-learning method based on cognitive conflict has the following problems: (i) too much time is spent on a few concepts that don't result in the construction of scientific concepts; (ii) explanations cannot be generalised to diverse phenomena; and (iii) students face difficulties in recognising and experiencing conflicts.

As for the teaching-learning method based on analogies, Stavy (1991) found that students do not express their ideas explicitly, nor do they need to be aware of the conflict, but they understand the analogous situations and do not run the risk of losing their self-confidence or opting for the wrong ideas. Thus, Mortimer & Scott (2002) consider teaching - learning by analogies as the alternative model of constructivism based on conceptual change, where students' conceptions cohabit with newly acquired scientific conceptions and are not replaced.

As for the teaching-learning process based on mathematical modelling, Porfírio & Abrantes (1999) found that it is allied to teaching-learning by analogies, with the further fact that mathematical investigations in the classroom effectively overcome misconceptions with the help of analogies.

According to the above and the way they fit into the context of this work, teaching and learning methods using analogies and mathematical modelling, as these methods give students the chance to check their cognitive errors by comparing their pre-conceptions with scientific conceptions, allow students to reflect on and decide about certain conceptions. From the same perspective, D'Ambrosio,

> He says that in view of this procedure described, it can be seen that a mathematical modelling activity encompasses more than studying mathematical content, but also the development of competences and the actual doing of modelling, i.e. the creation of models for dynamic learning, (D'Ambrosio 2015:44).

Given the importance of students' conceptions in the construction of knowledge, Coll (1988) recommends that the selection of programme topics in a mathematics curriculum should result from a comparison between the objectives defined for each level of education and the evidence from research into students' conceptions. We therefore believe that an appropriate selection of methods based on mathematical modelling in the constructivist approach necessarily involves combining the results of research into pupils' conceptions with assessment practices in a constructivist paradigm, where processes, attitudes and values are also assessed, as they are recognised as fundamental to the construction of knowledge and pupils' education.

6. Conclusion

The literature review carried out for this study indicates that mathematics education has been built up historically, taking on different meanings and transformations to construct itself as a domain of knowledge and domains of practice that brings together, studies and explains a set of issues relating to the teaching of mathematics, teacher training and research into the themes of mathematical modelling in the constructivist approach. Mathematical modelling in the constructivist approach can contribute to Mozambican schools in changing pedagogical practices in the teaching and learning of mathematics in the classroom, through problem solving as a starting point and as a pedagogical resource aimed at building knowledge, relating extra-mathematical reality to school mathematics, linking theory and practice, as this develops students' cognitive capacity and skills.

From an empirical point of view, the development of the study involves planning and carrying out training activities with teachers that include 1) reflecting on their own practices and on their own and their students' main difficulties in the classroom; 2) the experience and discussion of mathematical problem-solving activities that address non-mathematical contexts or genuinely mathematical contexts, identifying their pedagogical potential with the students; and 3) in accordance with the objectives of the curriculum proposed for the secondary cycle, selecting and developing modelling activities with students in this cycle, observing their relevance, as well as the interest, mathematical notions, hypotheses and different strategies mobilised by the students. In this last aspect, the presence of the teacher will be very important to manage the dynamics of the class, the moments of interaction, the confrontation and discussion of these hypotheses, strategies, difficulties and individual or group results.

Exercises that focus exclusively on mathematical procedures which, when proposed, suggest that students mechanically apply knowledge of algorithms and procedures that they have already learnt and are easy to apply,

EXAMPLE 1:

1) Calculate the roots of the following quadratic equation: $x^2 - 14x + 24 = 0$ or

$$x^2 - 14x + 24 = 0$$

$$x_{1,2} = \frac{-b \pm \sqrt{b^2 - 4ac}}{2a}$$

$$x_1 = \frac{14 + \sqrt{(-14)^2 - 4.1.24}}{2} = \frac{14 + \sqrt{196 - 96}}{2} = \frac{14 + \sqrt{100}}{2} = \frac{14 + 10}{2} = \frac{24}{2} = 12$$

$$x_2 = \frac{14 - \sqrt{(-14)^2 - 4.1.24}}{2} = \frac{14 - \sqrt{196 - 96}}{2} = \frac{14 - \sqrt{100}}{2} = \frac{14 - 10}{2} = \frac{4}{2} = 2$$

$Sol: \{2,12\}$

When posed, problems suggest that students search, investigate, use interaction, deepen their knowledge and previous experiences and develop a strategy for solving them. In this way, other activities can be alternated involving situations in which the students' different skills are valued, with these activities being broader (bringing in non-mathematical or genuinely mathematical contexts, based on a theme suggested by the students or proposed by the teacher).

EXAMPLE 2:

Grandma Mutatela owns a rectangular piece of land in Muhalazi measuring 22m by 30m. She wants to use part of one of the sides, which measures 22m, to make a rectangular enclosure measuring $48m^2$, to store the crops from her field. Is it possible to build this enclosure using 28 metres of canvas? What are the measurements of its sides?

Although when equating this problem we get the same equation as in the previous example, students find it difficult to solve this problem related to everyday situations.

$$Para \ x = 2 \ \Rightarrow \ x(28 - 2x) = 48m^2 \ \Leftrightarrow \ 2m(28m - 2.2m) = 48m^2 \ \Leftrightarrow$$
$$\Leftrightarrow 2m(28m - 4m) = 48m^2 \ \Leftrightarrow \ 2m.24m \ = \ 48m^2 \ \Leftrightarrow 48m^2 = 48m^2$$

$$Para \ x = 12 \ \Rightarrow \ x(28 - 2.x) = 48m^2 \ \Leftrightarrow 12m(28m - 2.12m) = 48m^2 \ \Leftrightarrow$$
$$\Leftrightarrow 12m(28m - 2.12m) = 48m^2 \ \Leftrightarrow 12m.4m = 48m^2 \ \Leftrightarrow \ 48m^2 \ = 48m^2$$

$$Solução : \{12\}$$

Note: Students should realise that the solutions to the equation are not always the solutions to the problem. They need to analyse and decide which of them fits the requirements of the problem

7.   Bibliographical references

BARBOSA, J. C. *Students' practice in the mathematical modelling environment: the outline of a framework*. In: BARBOSA, J. C.; CALDEIRA, & A. D., ARAÚJO, J. L. (Org.). *Mathematical Modelling in Brazilian Mathematics Education: research and educational practices*. Recife: SBEM, 2007.

BASSANEZI, R.C. *Teaching and learning with mathematical modelling*. S.Paulo: Contexto, 2002.

CARRERA, António Carlos et al. *Diretrizes para a Licenciatura em Matemática*. BOLEMA, Rio Claro, 1991.

CHAGAS, E. M. P. de Figueiredo. *Maths Education in the Classroom: Problems and possible solutions*. São Paulo, S/d.

GAZETTA, Marineusa. *Modelling as a Mathematics Learning Strategy in Teacher Training Courses*. S.Paulo- UNESP: Rio Claro, 1989.

IMENES, L. M. P. *A study on the failure of maths teaching and learning*. Master's dissertation in Maths Education. UNESP: Rio Claro, 1989.

INDE/MINED. *Basic Education Curriculum Plan: Objectives, policy, structure, study plans and implementation strategies* . Maputo, INDE/MINED, 2003.

LIBÂNEO, José Carlos. *Didactics*. São Paulo: Cortez, 1998.

LOPES, A. R. L. Vieira & BORBA, M. de Carvalho. *Trends in Maths Education*. Roteiro, Santa Catarina: UNOESC, 1994.

MIGUEL, António, et al. *Mathematics Education: brief history, actions implemented and questions about its disciplinarisation*. Revista Brasileira de Educação, n° 7, 2004.

POLAK, Ymiracy Nascimento Souza (Orgs.). *Distance Education: Educational foundations and policies and their impact on distance education*. Distance Education Training Course - UNIREDE, Brazil: Curitiba MEC/ UFP, 2000.

SANCHIS, I.P. & MAHFOUD, M. *Constructivism: theoretical developments and in the field of education*. Rev. Eletrônica Educação, v.4, n.1, 2010.

SANTOS, Vinício de Macedo. *School mathematics, the student and the teacher: apparent paradoxes and polarisations under discussion. Cad. CEDES [online]. 2008, vol.28, n.74, pp.25-38* SCHEFFER, Nilce Fátima.

The encounter between Maths Education and Freinet's Pedagogy. Rio Claro: UNESP, 1995.

SIMON, M. A. *Reconstructing mathematics pedagogy from a constructivist perspective.* Journal for Research in Mathematics Education, vol. 26, n. 2, 1995.

TINOCO, Lúcia. *How and when a teacher is doing maths education.* BOLEMA, Rio Claro, 1991.

VIECILI, C. R. CONFORTIN. *Mathematical modelling: a proposal for teaching mathematics.* Dissertation (Master's in Maths Education), Pontifícia

Printed by Books on Demand GmbH, Norderstedt / Germany